THE US AIR FORCE IN ACTION

Percy Leed

Lerner Publications ◆ Minneapolis

Lerner Publications Company
An imprint of Lerner Publishing Group, Inc.
241 First Avenue North
Minneapolis, MN 55401 USA

For reading levels and more information, look up this title at www.lernerbooks.com.

Main body text set in ITC Franklin Gothic Std.
Typeface provided by Adobe Systems.

Editor: Brianna Kaiser **Designer:** Mary Ross

Library of Congress Cataloging-in-Publication Data

Names: Leed, Percy, 1968– author.
Title: The US Air Force in action / Percy Leed.
Description: Minneapolis : Lerner Publications, [2023] | Series: US military
    branches (Updog Books) | Includes bibliographical references and index. |
    Audience: Ages 8–11 | Audience: Grades 2–3 | Summary: "Since 1947, the air
    force has been an important branch of the US military. Learn about airmen's
    responsibilities, how they carry out missions, the vehicles they use, and more"—
    Provided by publisher.
Identifiers: LCCN 2021043602 (print) | LCCN 2021043603 (ebook) |
    ISBN 9781728458281 (library binding) | ISBN 9781728463599 (paperback) |
    ISBN 9781728462486 (ebook)
Subjects: LCSH: United States. Air Force—Juvenile literature.
Classification: LCC UG633 .L425 2023 (print) | LCC UG633 (ebook) |
    DDC 358.400973—dc23

LC record available at https://lccn.loc.gov/2021043602
LC ebook record available at https://lccn.loc.gov/2021043603

Manufactured in the United States of America
1-50857-50194-1/21/2022

TABLE OF CONTENTS

DIVE INTO THE AIR FORCE

An air force pilot drops food and water from a plane 30,000 feet (9,144 m) up.

Another air force member rescues people. These actions help people who lived through a flood.

The air force's mission is to keep the United States' skies safe.

TROOP:
a group of soldiers

The US Air Force became a part
of the army in 1907. It became its
own branch of the military in 1947.

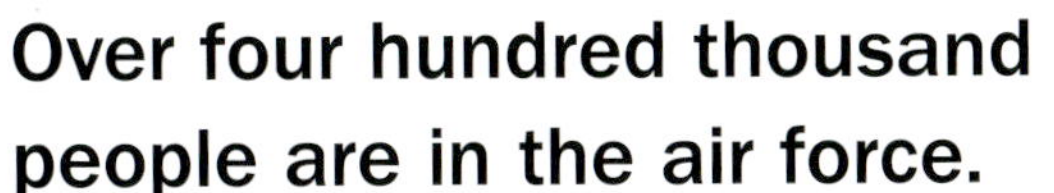

Over four hundred thousand people are in the air force.

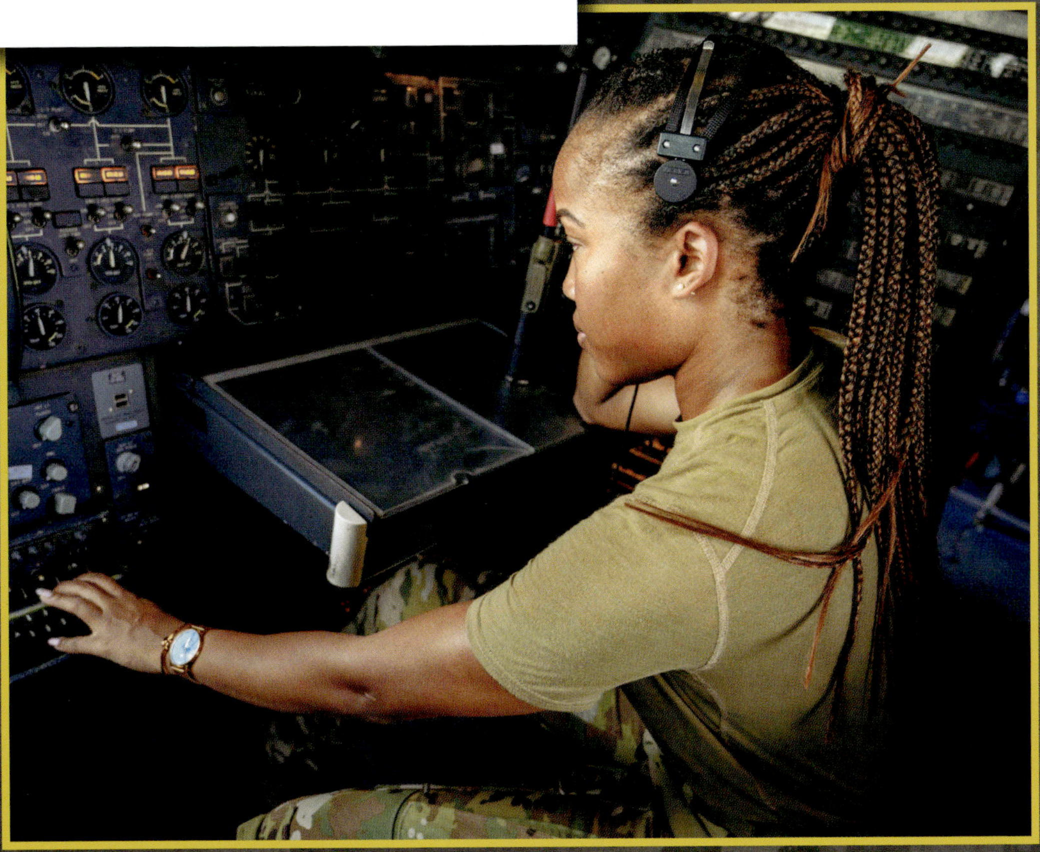

UP NEXT!

Forces on land.

ON THE GROUND

Many air force members work on the ground. Engineers design planes.

Air traffic controllers get pilots to where they need to go.

ENGINEER:
a person who designs
and builds things

VEHICLE CLOSE-UP

Air Force One is a plane that takes the president from place to place.

cockpit
wing
UNITED STATES OF AMERICA
engine
landing
gear

Air force members work on bases in the US, Europe, and Asia.

They often stay at a base for a few years before moving to another one.

BASE:
a place where air force members live and work

UP NEXT!

Up in the air.

FLYING HIGH

Air force teams fly people with medical emergencies.

They have a special plane set up with everything they need.

The air force expanded
into space in 1982.

In 2019, the air force space program became the US Space Force.

The air force works from the air and on the ground. It continues to grow.

MILITARY MISSION

FLOODWATERS RISE AND PEOPLE NEED HELP. WHAT COULD THE AIR FORCE DO?

A. Drop food and water

B. Fly people that need medical care to a hospital

C. Both A and B

Answer: C

GLOSSARY

base: a place where air force members live and work

engineer: a person who designs and builds things

troop: a group of soldiers

CHECK IT OUT!

Billings, Tanner. *The U.S. Air Force.* New York: Rosen, 2022.

Britannica Kids: Air Force
https://kids.britannica.com/students/article/air-force/272762

Ducksters: United Stated Armed Forces
https://www.ducksters.com/history/us_government/united
_states_armed_forces.php

Kiddle: Air Force Facts for Kids
https://kids.kiddle.co/Air_force

Murray, Julie. *Air Force Pararescue.* Edina, MN: Abdo Zoom, 2020.

Percy, Leed. *The US Space Force in Action.* Minneapolis: Lerner
Publications, 2023.

INDEX

PHOTO ACKNOWLEDGMENTS

Image credits: U.S. Air Force/Tech. Sgt. James L. Harper Jr., p. 4; U.S. Air Force/Tech. Sgt. Zachary Wolf, p. 5; U.S. Air Force/Tech. Sgt. Matthew Lotz, p. 6; U.S. Air Force/Senior Airman Mercedes Porter, p. 7; Library of Congress, p. 8; U.S. Air Force/Tech. Sgt. Michelle Y. Alvarez, p. 9; U.S. Air Force/Staff Sgt. Caleb Pavao, p. 10; U.S. Air Force/Senior Airman Breanna Klemm, p. 11; Pierre Albouy/AFP via Getty Images, p. 13; U.S. Air Force/Senior Airman Duncan C. Bevan, p. 14; U.S. Air Force/Airman 1st Class Donald Hudson, p. 15; U.S. Air Force/Staff Sgt. Nathanael Callon, p. 16; U.S. Air Force/Airman 1st Class Erin Baxter, p. 17; NASA, p. 18; Lockheed Martin Corporation, p. 19; U.S. Air National Guard/Tech. Sgt. Drew A. Egnoske, p. 20. Design elements: Andrey_Kuzmin/Shutterstock.com; anyababii/Getty Images.

Cover: DVID/Staff Sgt Jorrie Hart.